SpringerBriefs in Biotech Patents

Series Editor
Ulrich Storz, Duesseldorf, Germany

For further volumes:
http://www.springer.com/series/10239

Andreas Hübel · Thilo Schmelcher
Ulrich Storz

Biopatent Law: Patent Strategies and Patent Management

Springer

Dr. Andreas Hübel
Patent Attorneys
Michalski Huettermann
& Partner
Neuer Zollhof 2
40221 Duesseldorf
Germany
e-mail: hl@mhpatent.de

Dr. Ulrich Storz
Patent Attorneys
Michalski Huettermann
& Partner
Neuer Zollhof 2
40221 Duesseldorf
Germany
e-mail: st@mhpatent.de

Dipl.Ing. Thilo Schmelcher
Patent Attorneys
RCD-Patent
Karl-Heinz-Beckurts-Str. 13
D-52428 Juelich
Germany
e-mail: schmelcher@rcd-patent.de

ISSN 2192-9904
e-ISSN 2192-9912
ISBN 978-3-642-24845-0
e-ISBN 978-3-642-24846-7
DOI 10.1007/978-3-642-24846-7
Springer Heidelberg Dordrecht London New York

Library of Congress Control Number: 2011941757

Disclaimer: The information provided herein reflect the personal views and considerations of the authors. They do not represent legal counsel and should not be attributed to the companies or law firms the authors work for.

Printed on acid-free paper

Springer is part of Springer Science+Business Media (www.springer.com)

Preface to the Series

Biotech patents are a different world, even for patent practitioners who have obtained their expertise in neighbouring disciplines, like chemistry. One reason for this phenomenon is that, until about 20 years ago, novel biological embodiments were generally excluded from patentability. Classical breeding methods used for their creation relied on the random distribution of genetic matter, and thus lacked reproducibility and, hence, technicity—a criterion which is, in most patent jurisdictions, considered as a *conditio sine qua non* to qualify for patent protection.

With the rise in biotechnological methods, such as restriction enzymes, PCR, transfection methods and the like, a molecular toolbox is now available for the artisan which guarantees reproducibility with a sufficiently high percentage. Patent applications related to these methods therefore comprise a clear technical teaching. For current methods in biotechnology, technicity is thus no longer denied.

Biotech inventions are, however, facing headwind from another direction, too. Many biotech inventions are under public scrutiny for moral issues, or because they are considered as mere discoveries rather than inventions. Some countries have already established exclusions from patentability with respect to particular fields of biotechnology, or are about to do so. Argentinia has, for example, excluded genetically engineered plants,[1] while in the member states of the European Union, human embryonic stem cells are excluded from patent protection in the future.[2] A recent decision by the U.S. Court of Appeals for the Federal Circuit[3] has dispelled fears that gene sequences used for diagnostic purposes or therapeutic proteins isolated from nature would no longer be patentable.

At the same time, biotech inventions often require large investments in R&D, and can develop tremendous commercial potential, thus making patent protection a

[1] Arts. 6 + 7 of the Argentine Patent Act and Argentine Guidelines for Examination of Patent application, Part C, Chapter IV, 2.1.7.

[2] Decision of the European Court of Justice, case C-34/10, published on the website of the European Court of Justice (http://curia.europa.eu).

[3] Association for Molecular Pathology, et al. v. USPTO, Myriad Genetics, et al. v. Myriad Genetics, Inc. See No. 2010-1406 (Fed. Cir. July 29, 2011).

must to recover the invested resources. It is thus not surprising that the world of Biotech patents is a quickly developing one, which is sometimes hard to keep pace with.

Although plenty of literature related to Biotech patents is available, like Hans-Rainer Jaenichen's formidable book "From Clones to Claims",[4] a comprehensive treatise addressing the different Biopatent issues from all perspectives does not yet exist.

The present series "SpringerBriefs in Biotech Patents" tries to meet this goal. Each volume comprises three chapters devoted to three related topics, which are written by genuine experts of their discipline.

It is our hope that this series will help to create a better understanding of Biopatent issues, and support patent professionals to navigate the shallow waters of Biotech patents.

Duesseldorf, Germany Ulrich Storz

---

[4] Jaenichen HR, McDonell L, Haley J F, Hosoda Y: From Clones to Claims: The European Patent Offices's Case Law on the Patentability of Biotechnology Inventions in Comparison to the United States and Japanese Practice, Heymanns 2006.

About the Series Editor

Dr. Ulrich Storz was born in 1969 in Muenster. He graduated in Biology form the University of Muenster in 1998, where he received his Ph.D. in 2002. He is author and co-author of several scientific publications in the field of biology and biophysics as well as of several juridical publications in the field of intellectual property.

He passed the German Patent Bar Examination in 2005. Since 2005, he has been admitted to practice as European Trademark Attorney at the European Trademark Office (OHIM). In 2006, he was registered in the list of representatives before the European Patent Office.

This main practice areas in the field of Intellectual Property Law include Patent Prosecution, FTO and Patent Infringement, as well as Patent strategies; especially in the Life Science field (i.e. Biotechnology, Biophysics, Biochemistry, microbiology). One of his major fields of interest is Antibody IP.

Ulrich Storz is active as a speaker for the congress management company "Forum Institut für Management GmbH", and he organizes the annual Rhineland Biopatent Forum.

Preface to the Volume

Patents protecting biotechnological invention are becoming ever more important. Because biotechnology has many differences with respect to other technologies, lessons learned in other fields of technology cannot simply be transferred to adopt a suitable strategy for dealing with biotechnology inventions.

In this issue, general aspects of biopatent law will be discussed. This involves questions of patentability, including ethical issues and issues of technicity, as well as questions of patent exhaustion in cases where reproducible subject matter, like cells or seeds, is protected.

Another topic is active and passive patent strategies. Companies should not only develop an own patent portfolio, but should also monitor their competitor's activities, in order to be able to counteract should the need arise.

Further, insight will be given into patent lifetime management and additional protective measures, like supplementary protection certificates and data exclusivity. Here, strategies are discussed as to how market exclusivity can be extended as long as possible, which is particularly important for biopharmaceutical drugs, which create high R&D costs.

Duesseldorf, Germany

Ulrich Storz
Andreas Hübel
Thilo Schmelcher

Contents

General Issues of Biotech Patents

Andreas Hübel

Abstract Biotechnological inventions are clearly patentable when meeting the requirements set by the patent law. These requirements are based on well-established requirements. However, unique requirements were established such that the peculiarities of biotechnological inventions can be considered. Nonetheless, it appears that certain specific regulations for obtaining patent protection for a biotechnological invention poses more burden to an applicant than seeking patent protection in other disciplines.

Keywords Biotech · IP · Patents · Patentability · Invention · Novelty

1 Introduction

Humans have been using microorganisms and their metabolic products for centuries, for example when preparing food such as bread made of sour dough, cheese, wine, or other alcoholic beverages. However, it was only discovered in the mid nineteenth century that living microorganisms were responsible for such processes, and subsequently individual strains of microbes were isolated for further use. Besides, biology was a rather descriptive discipline sharing little with chemistry, physics, or engineering. Experimental biology such as the research conducted by Gregor Mendel (1822–1884) concerning genetic inheritance appeared to be an exception rather than the common approach. However, upon

A. Hübel (✉)
Patent Attorneys, Michalski Huettermann and Partner, Neuer Zollhof 2, 40221, Duesseldorf, Germany
e-mail: hl@mhpatent.de

A. Hübel et al., *Biopatent Law: Patent Strategies and Patent Management*,
SpringerBriefs in Biotech Patents, DOI: 10.1007/978-3-642-24846-7_1,

elucidating the double helix structure of DNA, characterizing the basic principles of how genetic information is passed and realized, i.e. the genetic code and messenger RNA, and identifying restriction endonucleases biological science evolved to a technical discipline nowadays called biotechnology.

Today, there is no doubt that molecular biology, molecular genetics, and biotechnology are disciplines that provide technical improvements and solutions for technical problems. Hence, the results provided by these disciplines are vulnerable to patent protection such that the inventors can commercialize their finding under a limited exclusivity. The latter appears to be particularly appropriate because bringing a biotechnological invention into the market is often a very developmentally intense project.

Patents are recognized as means for protecting intellectual property and inventions. Patents provide exclusivity for the inventor in commercializing his invention. By granting a patent, a state cedes some of its rights to the patentee. To allow a patentee to enforce the right that was ceded to him, the patent has to meet various criteria that are examined before the patent right can in fact be granted to the patentee. Over time those criteria became well established for technical disciplines and evolved along with the technical progress in the technical disciplines. Of course, these criteria were and have to be applied to biotechnological inventions as well. However, it turned out that some of the criteria were insufficient for appropriately protecting biotechnological inventions due to the peculiarities of this discipline. Hence, the criteria were amended, common criteria were interpreted in new ways when applied to biotechnological inventions, and new criteria were set up. Moreover, the intellectual property right for biotechnological inventions is still changing as ever new issues arise in this fairly young subject.

This chapter will illustrate and discuss the general issues of patenting biotechnological inventions, and emphasize the peculiarities in this regard. A focus is set on European patent regulations, but differences in other jurisdictions are mentioned if considered necessary by the author. All the decisions identified in this chapter can be accessed online via the database of the technical board decisions of the European Patent Office (EPO).[1]

2 General Principles

Patent law developed criteria for assessing whether an invention is patentable or not. These criteria are applied to biotechnological inventions too. These criteria are discussed and illustrated with respect to biotechnological inventions herein below.

[1] Available online under http://www.epo.org/law-practice/case-law-appeals/search_de.html.

2.1 Discovery Versus Invention

Patents shall be granted to inventions provided they are new, involve an inventive step, and are susceptible to industrial applicability. This is what basically every patent law in the world provides by one wording or another. Surprisingly, no patent law comprises a clear and unequivocal definition of what an invention in fact is. One has to rely on case law to figure out what—at least the German Supreme Court in the case "Rote Taube"[2]—regards as an invention. The red dove is a new species of doves resulting from crossing and selective breeding of already existing species. Although genetic crosses are biological processes, human selection is a technical aspect. Hence, the German Supreme Court concluded that a selective and systematic exploitation of natural forces including biological forces should not be excluded from patent protection.

The patent laws themselves do not define "invention". The patent laws may define what shall not be regarded as an invention. The EPC for example defines that discoveries, scientific theories, mathematical methods, aesthetic creations, schemes, rules and methods for performing mental acts, playing games or doing business, programs for computers, and presentation of information shall not be regarded an invention.[3] Hence, no European patent can be obtained for any of the said subjects. For example, it is a discovery if a natural ligand for a given receptor is identified. Hence, the binding of the ligand to the receptor is not patentable. However, a medicament comprising the ligand for treating a specific disease is patentable.

In addition, it is provided that European patents shall not be granted for inventions, the commercial exploitation of which would be contrary to "ordre public" or morality, plants, or animal varieties, essentially biological processes for the production of plants or animals, and methods for treating the human or animal body by surgery or therapy, as well as diagnostic methods practiced on the human or animal body.[4] One such example are human embryonic stem cells which require destruction of a blastocyst which in turn is considered as "human embryo."[5]

2.2 Novelty

For an invention to be patentable, it has to be novel. To be novel in the meaning of patent laws means that the invention shall not have been disclosed in any way prior to the first filing of the application for a patent. The invention shall not form part of the state of the art, i.e. the knowledge to mankind. State of the art in turn is everything made available to the public before the date of filing the patent

[2] GRUR 1969, 672.

[3] Article 52(1) EPC.

[4] Article 53 EPC.

[5] Decision G 2/06 of Enlarged Board of Appeals of EPA, Decision of ECJ in case C-34/10.

application. It does not matter at all, who made something available to the public (the applicant, the inventor, or any third party), and how it was made public (written description, oral presentation, use, or any other way). Thus, pursuant to the European understanding, novelty is absolute. The sole question is how to prove that the invention is novel or not.

While examining a patent application, the patent office conducts a literature search. In other disciplines, the vast majority of knowledge is present in published patent applications and patents. Hence, hardly any other literature than published patent applications are cited by a patent office when the said patent office questions novelty of the claimed invention, and/or its inventiveness. For mainly historical reasons, a lot of general information and knowledge concerning biotechnological information has been published in scientific journals. Hence, much more knowledge pertaining to biotechnology is present in other documents than in published patent applications compared to other technical disciplines. Thus, all parties involved in patenting a biotechnological invention are much more concerned with scientific literature than the parties involved in patenting other inventions. Nonetheless, a claimed invention is deemed to lack novelty if said invention is disclosed in a single document[6] in a direct, unambiguous and explicit manner. The claimed invention may to some extent also be disclosed in an implicit manner. When an explicit cross reference is made in a single document to another document, the disclosure is also considered as novelty destroying. Hence, scientific literature has to be considered routinely if patentability of a biotechnological invention shall be assessed prior to filing a patent application. Moreover, it is necessary for a company who wants to protect its IP by patents to have any other manuscript, in particular manuscripts for scientific journals which include sufficient technical details, reviewed for their content, and to assess whether the technical details disclosed therein would affect a patent application. Thus, filing and publication of (i) patent applications and (ii) scientific articles have to be coordinated to avoid disadvantages for the applicant.

To prove that a claimed invention is not novel is fairly easy, if the invention is disclosed in a single written document. However, oral descriptions such as presentations on conferences will destroy the novelty of a claimed invention in a patent application as well, if the oral presentation is made prior to the filing date of the patent application. Since it is more difficult to prove whether an oral presentation is novelty destroying for a claimed invention, it has to be established (i) at which date the alleged oral disclosure occurred, (ii) exactly what was said, and (iii) under which circumstances the alleged oral disclosure occurred. In this regard, it is to be noted that making a disclosure publicly available does not mean that the entire public must have been notified. It is sufficient if a non-restricted group of persons had the chance of becoming informed about the invention. It is not necessary to prove that a certain number of persons were indeed informed. Anyhow, establishing the above-mentioned criteria is not an easy task. But nonetheless,

---

[6] EPO Technical Board decision T 466/96, EPO Board of appeal decisions database.

inventors and applicants should be cautious about what to present at an oral presentation when they want to file a patent application covering the invention but not having done yet.

As an example with respect to novelty, a claimed nucleic acid molecule is novel if its sequence has not been published prior to the filing of the patent application claiming a nucleic acid molecule having a particular nucleotide sequence. This includes that the nucleic acid molecule should not have been deposited in a database from where it could be retrieved by an undisclosed number of persons. A claimed nucleic acid molecule whose nucleotide sequence deviates from the nucleotide sequence of a known nucleic acid molecule renders the claims of nucleic acid molecule novel over the known nucleic acid molecule. It does not matter whether the deviation is the absence of a single nucleotide at one of the nucleic acid's ends, the presence of a single additional nucleotide at the 5′ end or 3′ end of the nucleic acid molecule, or a nucleotide substitution.

The absolute novelty is not absolute, because a couple of exceptions exist, which prevent that disclosing an invention is novelty destroying for the subject matter of a subsequently filed patent application concerning the said invention. One such exception is the presentation of the invention on particular, official exhibition. Another example is the misuse of the invention by a third party who filed a patent application concerning the invention the said third party is not entitled to.

2.3 *Inventive Step*

An invention has to involve an inventive step in order to be patentable. Involving an inventive step means that the invention was not obvious to a person skilled in the art at the time the claimed invention was filed for obtaining a patent. This legal provision shall prevent—for example—simple alternatives from becoming patented. Thus, the requirement of inventive step can be considered as a qualitative measure preventing the patent system from being clogged by trivialities.

Whether or not a claimed invention is based on an inventive step is by far more difficult to judge than whether it is novel, because the answer to that question is at least to some extent subjective. For example, it has to be considered who the person skilled in the art is and what knowledge the person may have. The more knowledgeable a person skilled in the art is considered, the more inventions are likely to be deemed obvious. For the field of biotechnology, the consensus skilled artisan is not a Nobel Prize laureate,[7] not a highly skilled laboratory technician[8] nor the inventor.[9] The skilled artisan would consider means that were successfully

---

[7] EPO Technical Board decision T 60/89, EPO Board of appeal decisions database.

[8] EPO Technical Board decision T 412/93, EPO Board of appeal decisions database.

[9] EPO Technical Board decision T 5/81, EPO Board of appeal decisions database.

applied in a very close neighboring area,[10] but the skilled artisan does not possess any inventive capability[11] nor would he enter into unexplored areas[12] or even perform any scientific research in these areas.[13] Hence, the consensus skilled artisan in biotechnology is a team of cautious PhD bench molecular biologists including their laboratory assistants which are capable of practically applying methods in the art, which are aware of the disclosure of pertinent prior art documents, which have the necessary manual dexterity and lack of fatigue. However, they would not question established prejudices nor would they try to enter into sacrosanct of unpredictable areas.[14]

Attempting to render the examination of inventiveness more reliable and comprehensible, the EPO established the technical "problem–solution approach" including the "could/would approach". In this approach, the person skilled in the art is defined, and the closest prior art is identified. Usually the closest prior art is that single reference disclosing the combination of features which constitutes the most promising starting point for a development leading to the claimed invention. In doing so, the purpose is considered or the effect of the invention. Very often, the closest prior art reference is the reference disclosing the maximum number of technical features of the claimed invention compared to all other prior art references. Subsequently, the difference of the claimed invention and the prior art in terms of structural or functional features is identified, and the technical effect resulting from the distinguishing features are assessed which in turn is used to formulate the technical problem that shall be solved by the claimed invention. This technical problem is the "objective technical problem" and may be different from the technical problem the inventor wanted to have solved when he made his invention. Then the question to be answered is whether there is any teaching in the prior art as a whole, i.e. not necessarily in the same document, but in any other prior art document that would have prompted the skilled person to modify the known closest prior art to arrive at the claimed solution. This approach recognized that it is not sufficient that the prior art could have prompted the skilled person, but beyond that has to comprise some information which would have motivated the skilled artisan to do so. Practically, this means that the combination of two or at the utmost three prior art documents has to disclose all features of the claimed invention, and that at least one of the prior art documents has to include a suggestion for combining the technical features with a reasonable expectation of success. Whether an invention is based on an inventive step is a decision on a case-by-case basis.

[10] EPO Technical Board decision T 455/91, T 387/94, EPO Board of appeal decisions database.

[11] EPO Technical Board decision T 39/93, EPO Board of appeal decisions database.

[12] EPO Technical Board decision T 296/93, EPO Board of appeal decisions database.

[13] EPO Technical Board decision T 500/91, EPO Board of appeal decisions database.

[14] EPO Technical Board decision T 455/91, EPO Board of appeal decisions database.

2.4 Industrial Applicability

Legal provisions provide that an invention should have industrial applicability to be patentable. This requirement is rather broad and does not possess an undue burden, because an invention is considered to be susceptible of industrial applicability if the invention can be made or used in any kind of industry, including agriculture.[15] Biotechnology is an industry, and hence, inventions concerning or involving nucleic acids, proteins, microbiological strains, methods for producing (micro)organisms, and the like are susceptible of industrial applicability.

However, therapeutic and diagnostic methods were considered as lacking industrial applicability, because medical services were not considered to be an industry; all medical doctors should be free to choose the methods for treating their patients. Whether this presumption holds true remains to be discussed. However, Article 53(c) EPC provides that European patents shall not be granted for methods for treatment of the human or animal body by surgery or treatment and diagnostic methods plasticized on the human or animal body.[16] Hence, therapeutic and diagnostic methods as such are not patentable nowadays, regardless of whether they will be considered applicable in an industry or not.

The prohibition of patenting therapeutic or diagnostic methods does not apply to products for use in these methods.[17] Thus, biological molecules such as peptide hormones, antibodies, nucleic acid molecules which may be used in therapy can be patented.

For a sequence of a gene or a partial sequence of a gene, its industrial application has to be disclosed in the patent application.[18] Thereby, it is acknowledged in the EPC that even a partial sequence of a gene can have industrial applicability. However, the industrial applicability has to be disclosed in the patent application. Whether or not this disclosure requirement also affects the scope of a patent claim concerning a nucleic acid molecule will be discussed herein below.

2.5 Unity

A patent application shall relate to one invention only, or to a group of inventions so linked as to form a single general inventive concept.[19] Such a single inventive concept can be seen in a common technical feature of the different subject matters that are claimed in an application. Although a nucleic acid molecule may encode a specific protein which in turn can fulfill a specific function, the nucleic acid

---

15 Article 57 EPC.

16 Article 53(c) EPC.

17 Article 53(c) EPC.

18 Rule 29(3) EPC.

19 Article 82 EPC.

molecule is not part of the polypeptide. Thus, it appears that a nucleic acid molecule and a polypeptide which is encoded by the nucleic acid molecule do not share a common technical feature. The polypeptide and the nucleic acid sequence encoding the polypeptide are two distinct molecules with different features and utilities. Hence, it appears that claims within a single application can be construed as lacking unity, if they are directed to a nucleic acid molecule encoding a polypeptide, and to the polypeptide encoded by the nucleic acid molecule when the claims are drafted too broadly.

The same consideration may apply to claims directed to a nucleic acid molecule which is characterized by different features such as (i) its nucleotides sequence, (ii) its hybridization properties, and (iii) its similarity to a given nucleic acid sequence. Although apparent from the skilled artisan that all these subject matters are linked to one another by a single technical concept, such a single technical concept may not be sufficient to support unity of the invention by means of patent law. In such cases, the applicant may file one or more divisional applications to pursue his interest in all embodiments of the invention. However, the applicant has to face significant costs for the divisional applications as all fees due for the parent application have to be paid for each divisional application too.

3 Specific Regulations

This section provides information about specific regulations which were introduced into patent law for addressing peculiarities of biotechnological inventions. Although these regulations are reasonable, it appears that fulfilling the requirements provided by these regulations require more attention by those seeking patent protection for their biotechnological invention, more effort, and have to be considered in strategic considerations.

3.1 Biological Material

When a biotechnological invention concerns biological material, it may be difficult to describe the biological material in a sufficiently clear and precise manner in writing, for example when the invention concerns microbes. Anyhow, the invention has to be disclosed in the patent application in a manner sufficiently clear and complete for it to be carried out by a person skilled in the art at the filing date or priority date.[20] To overcome the conflict of not being able to clearly describe an invention concerning microbial strains and the requirement of doing so, it is possible to deposit the biological material in order to supplement the written description of the invention. With respect to the EPC, the EPO is a member

[20] Article 83 EPC.

"state" of the Budapest Treaty; the formal requirements are set out in Rule 31 of the Implementing Regulations.

It is noteworthy that US patent practice allows depositing microorganisms before grant of the patent, whereas the EPC requires that the deposition has to be made by the European filing date at the latest. Thus, claiming priority of a US patent application which refers to biological material which was not deposited at the filing date of said US patent application in a subsequent European patent application affects the claimed priority.

From the date the European patent application has been published, the deposited biological material must be available to any person or at least to an expert who has been nominated by the requester.[21] The deposited biological material is part of the application's disclosure. As the application is made public 18 months after its priority date, such that its disclosure becomes available to the public, the deposited material should also become publicly available. Therefore, it is prescribed that the applicant identifies the designation of the biological material, the deposition number, date of deposition, and has to identify the institution where the deposition was made.

From the day, the patent application has been published, any one may request a sample of the biological material, provided that the requester confirmed that he will not make available the said biological material to others, and will use for experimental purposes only, until the application has been withdrawn, is deemed to be withdrawn, or the patent has lapsed in all states.

The applicant does not have to fear that anyone can request obtaining a sample of the biological material from the institution of deposition. The applicant can restrict the public availability of the deposited biological material in choosing the "expert solution". This means that the applicant can request that a sample of the biological material shall only be provided to an expert who has been named by the applicant. The technical expert may then be the one assessing whether the deposited biological material indeed has the features disclosed in the application.

3.2 Nucleotide Sequences/Amino Acid Sequences

For inventions concerning nucleic acids or polypeptides, an applicant has to file a sequence listing which constitutes part of the application's disclosure.[22] The sequence listing has to be filed in written and a copy of the sequence listing has to be provided in computer-readable format also, such that the nucleotide sequences and amino acid sequences can be deposited in a database, and can be used for searching databases by the patent office to figure out, whether the relevant sequences are novel.

The sequence listing has to meet multiple formal requirements which render a sequence listing error prone such that it has to be corrected during the subsequent

---

[21] Rule 33(1) EPC

[22] Rule 30 EPC

prosecution. For the applicants' convenience the software PatentIn is provided, and can be downloaded from the internet pages of—for example—the EPO and the USPTO. An alternative software (BiSSAP) is available from the EPO too. The use of this computer program for preparing the sequence listing is highly recommended to reduce errors.

It became necessary for an applicant to identify the utility of the claimed nucleic acid in the description of the application. Back in the early days of genome sequencing, a tremendous number of ESTs were identified, and the nucleotide sequences were claimed in applications although the applicant did not have a clue to what protein an EST encodes and what the function of the said protein is. Hence, having an EST identified is rather a discovery than an invention. To render an EST becoming an invention, patent law was amended in that it became mandatory to identify the utility of the nucleic acid in the application text.

Interestingly enough, recent opinions demonstrate that the patenting of nucleic acids becomes more and more restricted to purpose-bound protection. One example is the recent ruling issued by the European Court of Justice (ECJ) against Monsanto, in which it was decided that patent claims related to genetic sequences are restricted, in their scope, to those embodiments in which the genetic information performs its function. The ECJ deemed that this was not the case for refined soybean meal comprising a claimed DNA providing herbicide resistance to the growing soybean plant.[23]

3.3 Stem Cells

Stem cells are a hot topic, not just in medical sciences, but in ethic discussions as well as with respect to intellectual property rights. With respect to intellectual property rights concerning pluripotent stem cells, the ECJ is concerned with the task of providing a definition of the human embryo for the needs of the protection of biotechnological inventions.[24] Article 6(1) of Directive 98/44/EC provides that inventions must be considered unpatentable where their commercial exploitation would be contrary to *ordre public* or morality. Article 6(2)(c) of the directive cites the use of human embryos for industrial or commercial purposes as an example of inventions which are considered unpatentable.

The ECJ considered that the concept of a human embryo applies from the fertilisation of an ovum by a sperm via the initial totipotent cells and to the entire ensuing process of the development and formation of the human body. As totipotent cells represent the first stage of the human body which they will become, they have to be legally categorised as embryos. It is irrelevant whether that categorisation must be recognised from before or only after nidation and whether fertilization occurred

---

[23] Case C-428/08 (Monsanto vs. Cefetra), published on the website of the European Court of Justice (http://curia.europa.eu).

[24] Case C-34/10

naturally or in vitro. The blastocyst is the product of the totipotent cell's capacity for development at a certain moment. The blastocyst is therefore one stage in the development of the human body. Accordingly, a blastocyst has to be categorised as an embryo, like any stage before or after that development.

Embryonic stem cells are not capable of resuming the development of the human body when they have been removed from the blastocyst. As pluripotent cells they cannot lead to a complete human being. Thus, embryonic stem cells, taken in isolation, cannot be categorised as human embryos. However, when pluripotent stem cells are removed from the blastocyst, the blastocyst will be destroyed by the said removal. As the blastocyst itself is categorized as an embryo, the removal of the pluripotent stem sells destroys the embryo they are removed from.

The ECJ considers that an invention has to be excluded from patentability, where the application of the technical process for which the patent is filed necessitates the prior destruction of human embryos or their use as base material, even if the description of that process does not contain any reference to the use of human embryos.

The proposal of the Advocate General is in line with the decision G 2/06 ("Stem cells/WARF") issued by the Enlarged Board of Appeal of the EPO, wherein the board ruled that claims directed to products which could be prepared exclusively by a method which necessarily involved the destruction of the human embryos from which the products are derived is excluded from patentability under EPC, even if the method is not part of the claims.

Thus, totipotent stem cells will not be patentable in Europe. Pluripotent stem cells may be patentable if they were not obtained from a blastocyst by destructing an embryo.

3.4 Antibodies

Antibodies are the fastest growing group of protein therapeutics. Hence, antibodies are among the molecules most often found in patent applications in Biotechnology. The desired protection may involve the antibody itself and the methods of generating/producing the antibodies. For the former, claims are drafted wherein the antibody is characterized by its amino acid sequence, by its binding properties, or by the deposition of the cell line producing the antibody. Antibodies may also be claimed by the process of how they are made.

Some kind of protection may also be obtained for already existing antibodies, for example by claiming a different medical use of the antibody than known, or as means of combination therapy with another compound. In the latter case superior efficacy has to be proven by the applicant in order to demonstrate inventiveness.

However, as antibody generation and selection processes become more and more straightforward and automatized, patent prosecution for antibodies thus produced will become more difficult.

3.5 Breeding Technologies for Plants

The European Patent Convention provides that no patent shall be granted for "essentially biological processes for the production of animals or plants". However, it has not been defined what an "essentially biological process" is, or what renders a process "essentially biological". These questions of law were referred to the enlarged Board of Appeals which answered that a non-microbiological process for the production of plants which contains or consists of the steps of sexually crossing the whole genomes of plants and of subsequently selecting plants is in principle excluded from patentability as being "essentially biological" within the meaning of Article 53(b) EPC. Such a process does not escape the exclusion of Article 53(b) EPC only because it contains, as a further step or as a part of any of the steps of crossing and selection, a step of a technical nature which serves to enable or assist the performance of the steps of sexually crossing the whole genomes of plants or of subsequently selecting plants. If, however, such a process contains within the steps of sexually crossing and selecting an additional step of a technical nature, which step by itself introduces a trait into the genome or modifies a trait into the genome or modifies a trait in the genome of the plant produced, so that the introduction or modification of the trait is not the result of the mixing of the genes of the plants chosen for sexual crossing, then the process is not excluded from patentability under Article 53(b) EPC. In the context of examining of whether such a process is excluded from patentability as being "essentially biological" within the meaning of Article 53(b) EPC, it is not relevant whether a step of a technical nature is a new or known measure, whether it is trivial or a fundamental alteration of a known process, whether it does or could occur in nature or whether the essence of the invention lies in it.

Providing this decision, the Enlarged Board of Appeal of the EPO provided clarity for applicants on what may be patentable, and on what is definitely not patentable as being considered to be an "essentially biological process". However, plants obtainable by an essentially biological process remain patentable if the remaining patentability requirements are met.

4 Conclusion

Biotechnological inventions are clearly patentable when meeting the requirements set by the patent law. These requirements are based on well-established requirements. However, unique requirements had to be established such that the peculiarities of biotechnological inventions can be considered. Nonetheless, it appears that certain specific regulations for obtaining patent protection for a biotechnological invention poses more burden to an applicant than seeking patent protection in other disciplines.

Active and Passive Patent Strategies

Thilo Schmelcher

Abstract This chapter deals with active and passive patent strategies. Several aspects will be highlighted from patent filing strategies towards litigation strategies both from an active viewpoint as claimant and a passive viewpoint as a respondent. Active and passive will be interpreted with respect to an intellectual property rights (IPR) portfolio and it will be interpreted in light of actions performed with respect to an IPR portfolio. Both views have an influence on each other. Even though these perspectives are at first hand seen as independent, typically a claimant will also become a respondent. Also the portfolio situations may change and so—up to a certain point in time—an IPR portfolio may still be shaped to same extent.

Keywords Patent · Prosecution · Infringement · Portfolio · Litigation

Pre-Remark In the following it is assumed that a product is based on certain technologies while a competing product may be directed to a like product and may be based on same technologies and/or other technologies. Also, we will refer to IPR portfolios, which encompass any kind of intellectual property right such as patents, utility models, supplementary protection certificates, trademarks, designs. Nevertheless this article will focus only on patents and utility models.

T. Schmelcher (✉)
Patent Attorneys, RCD-Patent, Karl-Heinz-Beckurts-Str. 13, D-52428, Juelich, Germany
e-mail: schmelcher@rcd-patent.de

A. Hübel et al., *Biopatent Law: Patent Strategies and Patent Management*,
SpringerBriefs in Biotech Patents, DOI: 10.1007/978-3-642-24846-7_2,

1 Introduction

Defining a strategy should always be the starting point when engaging in IPR. As many aspects are to be considered the strategy needs to reflect the competition. Competition thereby involves a strategic analysis of competing products, competing technologies as well as competing products. Once the competitive situation is understood it is possible to adopt a corresponding filing and/or litigation strategy which reflects the own ambitions as well as the competitive situation. In the following aspects of this process will be highlighted.

2 Portfolio View

Active and passive strategies for an intellectual property rights (IPR) portfolio may be more precisely understood as a proactive or a defensive filing strategy. Both strategies shape the initial IPR portfolio and influence portfolio maintenance costs. Also the question where (and why) to file shall be taken into account.

2.1 Defensive Filing Strategy

A defensive filing strategy is directed to own products and technologies tied to these own products. Within the defensive strategy the portfolio is only envisaging a possible protection of these products such as to be able to remove competitors from market for the patent term. Often, this strategy is backed up by thorough Freedom-to-operate studies to secure these products. In order to provide a proper basis for later-on attacks of competitors, creation of prior art is crucial. Hence, such a strategy should also envisage defensive publications and/or the filing of early published application documents such as a German utility model which is held to be public on the date of registration. Also in view of the so-called Hilmer doctrine,[1] filing of US Provisional Application is a necessity. This might change as the discussion within the US seems to provide a basis that the Hilmer doctrine will be abolished.[2] This type of strategy is often used in a

[1] Under this doctrine, named after decision "In re Hilmer 359 F.2d 859 (CCPA 1966)" it makes sense to submit a first filing before the USPTO to ensure that the application becomes prior art at the priority date (and thus prevent third parties who file a patent application for the same subject matter in the meantime in the US from obtaining a patent therefore), at least in case the text of the patent application is not in English.

[2] See "The Munich Talks: A Global Call for Harmonization", Under Secretary of Commerce for IP & Director for the USPTO David Kappos, http://www.uspto.gov/news/speeches/2011/kappos_munich.jsp.

later stage of product development if the value of a particular development is noticed.

2.2 Proactive Filing Strategy

In contrast, a proactive filing strategy is starting already in early stages of product development. A thorough analysis is performed to identify whether there are certain envisaged competitive products and whether these products use certain technologies and if there are competing technologies. Once these competitive products and the concerned technologies are identified a cluster of IPR applications are filed in order to prohibit competitors not only with respect to a particular product and a particular technologies but also to minimize the risk of competing products based on other technologies.

This typically leads to more complex portfolios having as a minimum the passive strategy while being supplemented by other aspects. This portfolios needs to be tracked in order not to spend too much money in areas which do not offer a proper protection. However, it will be hard to provide a fair estimation of avoided competition and therefore offered enhanced revenue streams. That is, the success of the proactive strategy may not be proven, which makes it sometimes hard to be communicated to business management.

There is a particular pro of this concept. That is, if a certain technology previously identified as a substitute emerges to a key technology for the product, than the proactive technology may still provide a proper basis such as to provide at least a defensive IPR portfolio for the product.

2.3 Where (and Why) to File

Having dealt with the question of what to file, the next important point is where to file and on why to file in particular states. Within this question several aspects needs to be highlighted. The most obvious point is to seek protection where products are produced or hit the market. Nevertheless, such an approach having a worldwide sold product would lead to oversized portfolios leading to high costs. Hence, one needs to identify boundary conditions.

2.4 Legal Stability and Absence of Corruption

A first boundary condition is legal stability and absence of corruption. One may not put large amounts of money in IPR prosecution in unstable legal systems. There are certain indices which may provide a basis for evaluation, namely the

Worldwide Governance Indicators published by the World Bank Group[3] and the Corruption Perception Index published by Transparency International.[4] However, there is a certain grey zone as legal systems (as well as political systems) are evolving. As such one needs to envisage future changes in legal systems. An example might be the People's Republic of China. While being renowned in the past for deficiencies in P law enforcement, and being suspected as having a certain favor for domestic parties, this air changes due to exchange programs for examiners and changes within the legal system. These projects are supported by several offices and in particular by the European patent Office.[5]

2.5 *Quality of Rulings*

Another boundary condition is the quality of rulings, i.e. whether there is an established jurisprudence having experience in IPR matters. There, one needs to identify those states which offer a certain number of new filed litigation cases per year, which can be seen as a tool to identify whether the court system is adopted to provide quality rulings since otherwise nobody would claim there.

2.6 *Other Boundaries*

Other boundaries pertain to certain circumstances necessitating a more broad application style. Such a circumstance may be the doctrine of exhaustion which may apply in a unitary market such as the European Union. That is, once, a product has been introduced within the economic community by a patent proprietor, it may circulate throughout the internal market irrespective of patent proprietors' IPR rights in other territories within the internal market.[6] This situation is of particular importance within Pharmaceutics and leads to a necessity for companies to have their products protected in any market within the internal market. Nevertheless if a new territory enters internal market by accession to the European Union the doctrine of equivalence may be subject to different transitional rules[7] and a filing

---

[3] http://info.worldbank.org/governance/wgi/index.asp

[4] http://www.transparency.org/policy_research/surveys_indices/cpi/2010

[5] http://www.ipr2.org/index.php?option=com_content&view=section&layout=blog&id=12&Itemid=89

[6] Federal Court of Justice "Bodenwaschanlage" X ZR 13/99, BGHZ 143, 268 - Karate, BGH, Urt. v. 24.09.1979 - KZR 14/78, GRUR 1980, 38, 39 - Fullplastverfahren.

[7] A transitional rule was part of accession act with respect to Czech Republic, Bulgaria, Estonia, Lithuania, Latvia, Slovenia, Hungary, Poland and Slovakia in 2004 respectively 2007, however no such rule was included for Cyprus and Malta.

strategy needs to envisage also these further accessions in order to have all future territories within the internal market protected.

3 Litigation View

Within Litigation view the focus will be on European Litigations. Both aspects will be highlighted, being a claimant and being a respondent. Although these aspects are embodied most often within a single party, it nevertheless will be examined independently.

3.1 European Patent System

A European patent provides a unified examination and grant procedure with respect to all member states of the European Union as well as the European Economic Area (European Union + Norway + Iceland + Liechtenstein) plus Switzerland, Turkey, Albania, Monaco, Former Yugoslav Republic of Macedonia, Serbia, San Marino and by way of so-called Extension Bosnia and Herzegovina and Montenegro.[8] Furthermore a European patent may be registered in Hong Kong.

Once granted, a European patent needs to be validated in the individual member states such as to become effective. This validation process requires in the majority of member states that the granted European patent is translated in one of the official languages.

3.2 London Agreement on the Application of Article 65

There is a certain relief for those having interest only in major litigation states as the London protocol removes the burden of translation in some member states.[9]

These member states are Germany, United Kingdom, France, Lichtenstein, Switzerland, Luxembourg, and Monaco.

Some member states to the London protocol require that only the claims are to be translated into an official language; these are Croatia, Denmark, Hungary, Iceland, Latvia, Lithuania, The Netherlands, Sweden and Slovakia.

Some of the later ones even dispense the translation requirement if the European patent has been granted in English. These dispensing states are Croatia, Denmark, Hungary, Iceland, The Netherlands and Sweden.

---

[8] For a Map, see http://documents.epo.org/projects/babylon/eponet.nsf/0/E65E85FAF2F200F4C125744A00294866/$File/epo_member_states_10_10.gif.

[9] For actual data, please consult http://www.epo.org/law-practice/legal-texts/london-agreement.html.

Compared to a plurality of national applications a European patent is more cost effective. Cost effectiveness may even increase in the near future as there seems to be a drive towards a "community" patent. This "community" patent will be a European patent validated for all member states of the European Union but Spain and Italy. According to recent documents being basis for consultation (KOM(2011) 216/3), the translation regime affords that in case of litigation, the patentee has to provide on request and to the choice of language of the respondent a complete translation whereby the language choice is restricted towards one of the official languages in which the alleged infringement took place or is domiciled. In addition a court may request a translation towards the official language which is used within the court procedure (Article 4). Additionally, it is foreseen that for a transitional period all patents needs to be translated into English if the language of procedure was German or French, and a translation of the patent as a whole towards another official language of one of the participating member states. The transitional period will last for at least 6 years and will then be reviewed (Article 6). Hence, the translation cost will be dramatically minimized as for the majority of cases only one translation will be necessary.

3.3 Litigation Statistics

Now turning to some statistics may provide another view on this issue which has significant impact, i.e. the question where today patentees are filing their infringement suits. An analysis[10] of this question reveals that the vast majority of litigations within the European Union (for the period 2003–2006) was handled in only a small number of states. These are Germany, United Kingdom, The Netherlands and France handling about 90% of all European patent litigations. These numbers are stable.[11] The majority of cases are handled by German courts based on a validated European patent while the numbers of infringement suits handled in the other countries is substantially lower. The reason for this distribution may be seen in the market size of the state itself (Germany, France, United Kingdom), the position within global trade (The Netherlands), the reliability of decisions and the legal system in general (Germany, France, United Kingdom, The Netherlands), the time to decision and availability of instruments for obtaining evidences. Most often decisions rendered in one of these states are in fact blueprints for other courts even though no court would admit so. Therefore, any litigation strategy should envisage these states.

---

[10] http://documents.epo.org/projects/babylon/eponet.nsf/0/2D620982152DB517C12572A70043 3C61/$File/impact_assessment_2006_02_v1_en.pdf

[11] Data shown in the context of a European Patent Litigation Insurance show said trend, see Figure 1 on page 110 of http://ec.europa.eu/internal_market/indprop/docs/patent/studies/pli_appendices_en.pdf.

3.4 Court Systems

In the following some main points within the respective legal systems will be highlighted.

3.4.1 Germany

In Germany there are 12 assigned patent litigation courts.[12] The majority of litigation cases (about 80%) are handled by the courts in Düsseldorf, Mannheim and Munich, whereby the order displays the respective ordering in descending manner. These courts are handling the cases by three legally qualified judges. Evidences may be obtained by means of special orders and are typically granted towards a court appointed technical expert. Typically, there is no immediate disclosure towards the claimant. The German court system is bifurcated, which means that invalidity of a claim may be responded in infringement proceedings but needs to be pursued within another specialized court, the Federal Patent Court. Infringement proceedings will stay only, if there are strongly convincing arguments towards invalidity of the patent and if an opposition (within 9 months after grant of a European patent or if another opposition is still pending) or a nullity action based on said arguments is actually filed. The Federal Patent Court handles nullity cases in separate proceedings but both courts take note of arguments made if they are made available to the court. The Federal Patent Court handles these nullity cases by two legally qualified judges and three technically qualified judges. The technically qualified judges are selected on the respective IPC classes the patent under attack bears. Within litigation and nullity, the losing party is obligated to pay the court fees as well as attorney fees within statutory limits based on a so-called value in dispute. If the claimant requests desist, it is still up to the claimant to decide when to enforce such an order, i.e. the claimant stays in control.

3.4.2 United Kingdom

The situation in the United Kingdom is rather different. There the legal system is spilt up into three territories. However, the major part of patent infringement is dealt with in England and Wales. But it is not only the territory which is split it is also the fact that there exist different paths to claim invalidity and infringement. Infringement may not only be claimed at the Patents County Court where a single judge is dealing with the claim and a possible counter-claim of invalidity but as well at the Intellectual Property Office if the parties agree to do so. Both patent court and Intellectual Property Office may transfer a case towards

[12] Düsseldorf, München, Mannheim, Berlin, Braunschweig, Erfurt, Frankfurt, Hamburg, Leipzig, Nürnberg, Magdeburg und Saarbrücken.

the Patents Court. It should be noted that the system necessitates barristers and solicitors and a patent attorney as technical expert thereby being extremely expensive. A pro to file a claim in England and Wales may be the availability of Anton Piller orders also known as search orders, by which a bailiff and a technical expert may take evidences. Depending on your claim language, e.g. if your claim pertains to "isolated gene sequences", it may no good idea to file a suite in the UK as a recent decision denied infringement if no isolated gene sequence is found in the alleged infringing product as sold. As there is a need within some of the above mentioned paths to have barristers being instructed by solicitors these proceedings very often get extremely expensive. It should also be noted that once enforcement is decided, the execution of the decision is out of claimant's control, which might be negative if one is about to achieve a settlement but still needs some time for finalization.

3.4.3 France

In France in the past there had been seven assigned courts dealing with patent infringement. Nowadays all cases are handled by the tribunaux de grande instance in Paris having 4 sections. There, three legally qualified judges are dealing the case. The court may appoint a technical expert. Before the Enforcement Directive of the European Union (Directive 2004/48/EC[13]) was put into practice France was often chosen as venue as it allowed for an instrument known as "saise contrefacon" which in fact is comparable to an Anton Piller order. This is a typical pre-trial instrument which is used in about 80% of cases. Once allowed, the claimant may even enter the respondent's premises. A saise contrefacon may even be obtained in respect of pending applications. After the actual "saise" the claimant has up to 20 working days respectively 31 calendar days whichever period last longer, to start infringement proceedings. So one needs to be rather specific on the type of information and one needs a very detailed knowledge where to search and should even then be prepared to start infringement proceedings rather quick.

3.4.4 The Netherlands

All patent litigations are handled by the district court in The Hague. There three legally qualified judges are dealing the case. The court may appoint technical experts. Nullity of a patent may be counterclaimed or claimed in defense. A specialty is accelerated proceeding, i.e. proceedings which typically last only 10–12 months. Within this accelerated regime the ping-pong of responses sent is strictly limited in time and number. Accelerated proceedings may be

[13] http://eur-lex.europa.eu/LexUriServ/LexUriServ.do?uri=CELEX:32004L0048R(01):EN:NOT

"downgraded" to normal proceedings, but normal proceedings may not be "upgraded". The Hague is again an interesting venue for cross-border litigations, as a recent referral[14] shows. The litigation costs compared to the size of market are relatively high.

3.5 Cost Dimension

Costs in relation to the respective market are relatively high in The United Kingdom and The Netherlands. Costs in relation to the respective market are in average relatively low in Germany.

According to a study prepared by the EPO in the course of discussion relating to the unfinished EPLA, the minimum costs of a small to medium-scale patent case at the second instance handled in the UK was about 150 T€ while the minimum cost in Germany was about 50 T€. At the same time the study indicated minimum costs of only 50 T€ for France and 60 T€ for The Netherlands. At the same time the maximum costs of the same small to medium-scale patent case was given to be 1500 T€ for a case handled in the United Kingdom and about 200 T€ for Germany, France and The Netherlands.

The minimum costs of a small to medium-scale patent case at the second instance handled in the UK was about 150 T€ while the minimum cost in Germany was about 90 T€. At the same time the study indicated minimum costs of only 40 T€ for France and The Netherlands. At the same time the maximum costs of the same small to medium-scale patent case was given to be 1000 T€ for a case handled in the United Kingdom and below 200 T€ for Germany, France and The Netherlands. Hence, compared to the market size the costs are quite low in Germany and the bandwidth is rather low providing a good cost estimate.

3.6 Claimant's Situation

Now turning back to actual litigation, at first we will assume the situation of a claimant. As a first point any claimant should check whether he believes that the IPR to be used is held valid.

For this reason, it is highly recommended to check all file histories within a given family and to compare all arguments as to whether documents cited in one legal system are valid prior art in another legal system. Beware of counter-part applications which might be unpublished but nevertheless be held as prior art at least for novelty attacks. State of the art definitions are sometimes different with respect to different types of IPR, e.g. the state of the art definition with respect to a

[14] C-616/10, Solvay vs Honeywell, referral of a question by the ditict court of The Hague, http://eur-lex.europa.eu/LexUriServ/LexUriServ.do?uri=OJ:C:2011:089:0009:0009:DE:PDF.

utility model in Germany is different to the state of the art definition with respect to a European or German patent.

3.6.1 Customization

Check whether there are still applications pending and whether these applications can be customized towards respondent's infringing acts. Customization may be achieved in several ways.

3.6.2 Divisional Applications

First of all, check whether there is still the possibility to file divisional European patent applications. A pending application in the background may also allow for customization later-on, once a respondent argues in a manner that a certain infringing act is not happening because …… Beware of the deadlines set out in Rule 36 EPC which are 24 months after the first communication under Article 94(3) EPC (i.e. an Examination Report) or Rule 71 EPC (e.g. Intention to grant) issued or 24 months if an objection is raised with respect to a particular lack of unity for the first time (Article 82 EPC).[15] Also if there are national applications pending, check whether it is still possible to file divisional applications.

3.6.3 Utility Models

Another possibility to customize IPR is to branch of utility models. Utility models by means of a branch of may be attained in Germany, Austria, Hungary and Czech Republic.

German utility models may not be directed towards methods, however, the Federal Court of Justice ruled in 2005 in decision "Arzneimittelgebrauchsmuster" (BGH X ZB 7/03) that a utility may be registered also for the use of a known substance within a second medical use.

In Austria, method claims are not excluded from patentability. Also, an Austrian utility model may claim chirurgical or therapeutical treatment of animal bodies while the same would be exempted within patents.

Hungary to the contrary requires that a utility model can only be granted for a form, structure or arrangement of components.

Czech utility models are not granted for methods.

Note, there exist different deadlines for branching of in every concerned member state and the calculation of deadlines is sometimes a mix of European and national deadlines. Even though one would at first think that a deadline lapsed,

[15] http://www.epo.org/law-practice/legal-texts/html/epc/2010/e/r36.html

it might happen that the filing of an opposition against a European patent reopens the deadline for filing utility models. Thereby, the situation might get tricky for the claimant as well as for the respondent. Suppose that the deadline for branching off has lapsed but the deadline for filing an opposition against the European patent is still running. Than the patentee may either choose to wait until the deadline for filing an opposition lapses before filing an infringement proceeding and thereby forcing the respondent to invalidate a granted European patent in several member states. This will bind resources, both in personnel as well as in respect to money. On the other hand, filing a claim early may trigger an opposition and reopen certain branching of periods which (as long as the term of a utility models is not lapsed) may be used by the patentee to customize utility models even in view of the responses given by the respondent.

This may be of particular relevance if prior art of the patentee is cited which would qualify for a 6 months grace period which is available with utility model regime. Also the national utility model provisions may exempt certain types of prior art, e.g. presentations abroad. Hence, utility models may turn into the only available instrument to attain IPR.

3.7 Respondent's Situation

Switching now to the respondents view. As a respondent one of the first steps is the evaluation of the claim and thereafter the evaluation of possibilities to get rid of an IPR by filing an opposition or a nullity. Knowing the prior art, it is necessary to evaluate whether the patentee may be brought into a better position by filing opposition. In that case, even though the costs for prosecution of nullity claims in several member states may be expensive, it may be worth the money if otherwise the patentee may be able to file customized utility mode or utility models may not subject to the prior art. In this connection it is recalled that decisions rendered in one jurisdiction often are blueprints for other jurisdictions.

But infringement is only one possibility to be active. Another is the continuous observation of competitors' portfolios and the attack via oppositions. Note, as oppositions are seen as grant related and as the opponent has to bear its own costs, oppositions are very often used to get patents revoked as early as possible to clear out any problematic IPR within a given field. With respect to nullity suits, there a certain cost risk is still given, as most jurisdictions stick to some sort of apportionment of costs to be borne by the losing party.

Again, an opponent-to-be or claimant to-be should consider whether an attack may reopen branching off possibilities thereby jeopardizing the effect intended to achieve and whether an action filed is not waking a dog which otherwise would have kept sleeping.

4 Multiparty View

Having identified several aspects within filing and litigation, it should be noted that all these aspects are mutually interrelated. It becomes even more tricky to find a good strategy if the situation of competitors is more complex.

Suppose the following situation, as a patentee a product P being based on a technology A and B is produced and subject to patent protection. A first competitor C1 has a competing product CP1 based on technology A and C while a second competitor has another competing product CP2 based on technologies B and C. In this situation prospective business data needs to be added to the picture to find out what it would mean if a certain technology comes under attack, i.e. weakening one competitor by attacking might impact another competitor by allowing the usage of a certain technology without fearing to be blocked or having to pay royalties. This may endanger the commercial success more than having the IPR around and seeking a settlement by means of a free license.

Hence, within multilateral situations, choosing the right strategy becomes more and more complex. These aspects should be taken into account in the beginning and needs to be reevaluated regularly. It is highly recommended to discuss any aspect not only within a patent department but to also share your thoughts with your outside counsels.

5 Summary

Active and passive strategies for filing have been highlighted. Several aspects impacting filing strategies such as legal stability, jurisprudence and cost effectiveness as well doctrine of exhaustion has been touched. Possible strategies for a claimant have been highlighted as well as a respondent's view was shown. As stated in the beginning, these viewpoints are depending on each other and in real life most claimants will become respondents and therefore will face like problems. Hence, starting litigation actions should be embedded in a plot envisaging counter actions to happen and to define an escalation strategy. Nevertheless, it should be emphasized that it is always easy to start actions but you should be aware and prepared to stop actions as well.

Patent Lifecycle Management, Supplementary Protection Certificates and Data Exclusivity in Biopharmaceutics

Ulrich Storz

Abstract The development of biopharmaceutics is a costly and time consuming matter. While patents compensate an innovator for the efforts and resources invested into research and development, their lifetime is restricted. However, due to lengthy authorisation processes for pharmaceutical products, biopharmaceutic innovators cannot exploit the full lifetime of their patents. In order to provide sufficient incentive to develop new drugs, all major countries have therefore developed exclusivity schemes related to regulatory procedures which complement the established patent protection systems. Further, patent lifecycle management strategies, which encompass the strategic filing of subsequent patent applications by the applicant, may help to extend the duration under in which a biopharmaceutic product can be sold under exclusivity.

Keywords Lifecycle · Patent · SPC · PTE · Biopharmaceutic

1 Introduction

The development of a new drug is a costly and time consuming matter. Today, the average costs to develop a new small molecular drug until it is ready for market entry are estimated to be about 800 million USD.[1] Despite earlier hopes, this situation has not changed for the better when biopharmaceutics, i.e. protein- or

[1] DiMasi et al. (2003).

U. Storz (✉)
Patent Attorneys, Michalski Huettermann and Partner, Neuer Zollhof 2, 40221, Duesseldorf, Germany
e-mail: st@mhpatent.de

A. Hübel et al., *Biopatent Law: Patent Strategies and Patent Management*, SpringerBriefs in Biotech Patents, DOI: 10.1007/978-3-642-24846-7_3,

nucleic acid-based drugs, were introduced. According to a study performed by the Tufts Center for the Study of Drug Development, the estimated average costs of developing a new biopharmaceutic are about 1.2 billion USD,[2] while development times are slightly longer than those reported for small molecular drugs.[3]

Because, usually, a patent application protecting a new drug has to be filed right at the beginning of drug development, patentees can exploit the exclusivity provided by their patent for a limited amount of time only, due to the long gap between patent application and final market entry. In order to provide sufficient incentive to develop new drugs, all major countries have therefore developed exclusivity schemes related to regulatory procedures which complement the established patent protection systems. The present chapter discusses both systems. Further, patent lifecycle management strategies, which encompass the strategic filing of subsequent patent applications by the applicant, will be discussed.

Although the present chapter focuses mainly on biopharmaceutics, some of the issues discussed (SPCs, Patent Lifecycle Management strategies) are also applicable to agricultural pesticides.

2 Patents and Related Monopoly Rights

Patents compensate an innovator for the efforts and resources invested into research and development. They provide an prohibitive right with respect to a given invention and thus affect free competition and exchange of goods. The lifetime of such privilege is therefore restricted, in order to provide public access to the subject matter of the invention once the patent has expired, and to enable third parties to use the once-protected technology.

2.1 Patent Lifetime

US Patents based on an application filed on or after June 8, 1995, expire 20 years after filing (35 USC §154(a) 2), with the option to extend the efficient protection period up to 21 years by claiming the priority of an earlier patent application of same or similar content, which is, in most cases, a provisional application filed at the United States Patent and Trademark Office (USPTO) or a foreign application filed in a member state of the Paris Convention.[4] However, a US Patent which is

[2] Average Cost to Develop a New Biotechnology Product Is $1.2 billion. Tufts Center for the Study of Drug Development, 9 Nov 2006. Available at http://csdd.tufts.edu/NewsEvents/NewsArticle.asp?newsid=69. Accessed on 27 October 2009.

[3] Grabowski et al. (2006).

[4] The Paris Convention is an international contract from 1883 on the mutual recognition of priority rights in intellectual property.

based on an application filed before June 8, 1995, expires either 20 years after the first US filing date or 17 years after grant, whichever ends later. This means that, for Patent applications filed before the said date, the effective protection period can be extended to more than 21 years. European Patents, in contrast, expire 20 years after filing [Article 63(1) EPC], which may amount to an effective period of 21 years as well in case a priority has been claimed. Similar terms apply to patents in most other industrialized countries.

2.2 *Patent Term Adjustment*

While there is no way to extend the actual lifetime of a European Patent, the USPTO grants day-for-day supplements to the normal patent term to compensate potential delays in prosecution caused by the USPTO under 35 USC 154 (b). These delays are offset against possible delays caused by the applicant. The process to calculate the actual patent term adjustment (PTA) considers predetermined time frames for both sides in the different stages of the patent prosecution; however, the term adjustment cannot be less than zero. The actual PTA is indicated on the title page of the published patent. The process is available for patents filed on, or after, May 29, 2000. The USPTO calculates PTA twice during patent prosecution using a computer algorithm, and informs the applicant accordingly. Due to the fact that a significant number of USPTO calculations bears mistakes,[5] applicants should consider to have a PTA calculation reviewed, particularly in case of compound patents protecting a biopharmaceutic, where each day of additional patent lifetime may result in substantial revenues. A qualified US patent counsel should be consulted, both to conduct the patent prosecution process in such way that highest possible patent term adjustment can be obtained, and to request, and achieve, a correction of the PTA should this be applicable.

2.3 *Patent Term Extension and Supplementary Protection Certificate*

Many countries have developed schemes to extend the protection granted by a drug patent in case the drug underwent time consuming approval proceedings.[6] Such term extensions play a paramount role in the protection of pharmaceutic

[5] Kayton (2008) Patent term duration and its calculation. Release No. 3, pp 54–55. Available at www.patetentterm.com.

[6] World Intellectual Property Organization: Survey on the grant and publication of "supplementary protection certificates of January 2002, available under http://www.wipo.int/standards/en/pdf/07-07-01.pdf.

drugs. An oft-cited example is Lilly's Prozac, patent protection of which expired, in Europe, in 1995. However, Lilly has achieved approximately 80% of their sales in the UK in the first 5 years after expiry of the Patent, i.e. when the product was under protection of a respective lifetime extension scheme only.[7]

In the US, the lifetime of the patent itself can be extended (so called Patent term extension, or PTE, not to be mixed up with Patent term adjustment) under 35 USC § 154 for a patent relating, among others, to the active ingredient of a drug product, a combination of such active ingredients, a medical device, or to methods of manufacturing or using such product, in case the product is subject to regulation under the Food, Drug and Cosmetic Act.

Patent term extension may be applicable in case the date of first marketing was delayed as a result of regulatory review, while the extension shall be for the same period as marketing was delayed as a result of the regulatory review. The respective term is calculated as one half of the term between grant of an investigational new drug exemption (which permits clinical testing in the United States) request for marketing authorization, plus the time said request is pending. The duration of the extension is restricted to a maximum of 5 years, yet the extension may not end later than 14 years after market authorization. Further, the scope of protection is limited to the product which has been authorized, an authorized use thereof or a method of manufacturing the authorized product depending on the claims of the patent.

In member states of the European Union (EU), compound protection for an approved drug can be extended under Article 63(2) of the European Patent Convention (EPC) and Regulation (EC) 469/2009 by means of a Supplementary Protection Certificate (SPC).[8] Contrary to PTE, an SPC is not an extension of the respective patent as such, but an exclusive right per se which refers to a given basic patent.

Under Article 7 of Regulation (EC) 469/2009, requests for SPCs must be filed within six months of the date of (1) the first market authorization for the said drug in any member state to the European Economic Area (EEA), i.e. EU + Iceland, Liechtenstein and Norway, or (2) grant of the patent, whichever ends later. Requests have to be filed at the different national patent authorities. The maximum lifetime of an SPC is 5 years, counted from the end of the patent term, and is calculated with the following formula:

$$\mathrm{LT} = \mathrm{MA} - \mathrm{FD} - 5\ \text{years}$$

(LT = Lifetime of the SPC, FD = Filing date of basic patent and MA = Date of 1st market authorization). This means that in case the market authorization has been granted less than 5 years after the filing date, the lifetime of an SPC is zero, or, theoretically, even negative.

---

[7] According to information provided by IMS Health Incorporated.

[8] For plant protection products, SPCs can be obtained under the rules set forth in regulation (EC) No 1610/96.

It is still unclear whether or not the scope of an SPC protecting a biopharmaceutic drug will also extend to a respective biosimilar, i.e. to a follow-on-biopharmaceutic which has undergone accelerated approval and is put to the market after patent expiry of the originator drug. Biosimilars have, for example, quite often a differing glycosylation pattern ("glycoform"). According to Article 4 of said regulation, the scope of an SPC extends "only to the product covered by the authorization to place the corresponding medicinal product on the market". In decision C-392/97,[9] the European Court of Justice (ECJ) has broadened said concept by stating that "where an active ingredient [...] is referred to in the marketing authorization concerned and is protected by a basic patent in force, the certificate is capable of covering [...] also its various derived forms such as salts and esters, in so far as they are covered by the protection of the basic patent". However, this decision relates to a small molecular drug and has probably not been made with biosimilars and their differing glycoforms in mind. Further, as biosimilars require a separate authorization—even if the latter were an abbreviated one—it is not sure yet whether or not they would fall under the meaning of regulation (EC) 469/2009.

One problem often encountered with SPCs is the correct match between (1) the product protected in the basic patent, (2) the product for which the first market authorization has been obtained and (3) the product which is subject of the SPC.

Like in the US, where the Court of Appeals for the Federal Circuit (CAFC) ruled that the issue is whether the patent actually claims the product that was subject to regulatory review,[10] quite a few requests have been rejected in the past due to failure to meet this requirement, as was the case for Erbitux in the UK, where two requests for SPCs (one for Erbitux alone and one for Erbitux plus chemotherapy) were eventually rejected as not being compliant with Article 3(b) of the regulation, according to which the product for which an SPC has been requested must be protected by a valid authorisation to place the product on the market (i.e. there needs to be a correct match between the product protected in the basic patent and the product which is subject of the SPC request). In this case, the basic patent related to Erbitux plus chemotherapy, whereas the marketing authorizations (one from Switzerland, and a later from the EMA) referred to Erbitux alone, while the combination with Irinotecan was mentioned as an option only. Accordingly, the court found no match between the basic patent and the authorized product.[11]

Interestingly, the case was decided differently in other member states of the European Union. In Belgium, Spain, France, Italy and Sweden only the SPC for Erbitux alone was granted, while in Austria only the SPC for the combination product was granted. In Greece and Luxemburg, both SPCs were granted, while in

---

[9] Case C-392/97 (Farmitalia Carlo Erba Srl's SPC application), published on the website of the European Court of Justice (http://curia.europa.eu).

[10] Decision Hoechst-Roussel Pharmaceuticals v. Lehman, 42 USPQ2d 1220 (Fed. Cir. 1997).

[11] Decision Yeda Research and Development Company Ltd v Comptroller General of Patents Decision, High Court, [82010] EWHC 1733 (Patent).

Table 1 An example for disharmony in SPC jurisdiction in Europe

Decisions	[82010] EWHC 1733 (Patent) (UK High court)
Basic patent	EP0667165B1, related to a therapeutic composition comprising a monoclonal antibody and an anti-neoplastic
First marketing authorization	CH: granted in respect of the medicinal product Erbitux-cetuximab, but description mentions use of Erbitux in combination with Irinotecan
Subject of SPC requests	SPC/GB 04/037 Erbitux + Irinotecan SPC/GB 04/038 Erbitux alone
Outcome	SPC/GB 04/037 rejected because market authorization was for Erbitux only SPC/GB 04/038 rejected because Erbitux alone was not subject of the patent
Member states where only the SPC for Erbitux was granted	Belgium, Spain, France, Italy and Sweden
Member states where only the SPC for Erbitux + chemotherapy was granted	Austria
Member states where both SPCs were granted	Greece and Luxemburg

the Netherlands, the judges decided similar as in the UK, i.e. both SPCs were rejected (see Table 1).

Obviously, the regulation (EC) 469/2009 seems to leave many other issues unsolved, too. On June 22, 2011, nine referrals were pending before the ECJ with respect to the interpretation of the regulation, out of which four relate Article 3(a) alone, according to which the product for which an SPC has been requested must be protected by a basic patent in force (i.e. there needs to be a correct match between the product protected in the basic patent and the product which is subject of the SPC request).[12]

Despite the legal uncertainty which obviously exists at least in the member states of the European Union, applicants should always ensure a correct match between the product of the basic patent and the authorized product. In case both a single agent drug and a combination drug are to be protected, applicants should seek individual patent protection for both embodiments, and care for individual market authorizations.

[12] Cases C-322/10 (Medeva BV v Comptroller-General of Patents), C-422/10 (Georgetown University, University of Rochester, Loyola University of Chicago v Comptroller-General of Patents, Designs and Trade Marks), C-630/10 (University of Queensland, CSL Ltd v Comptroller-General of Patents, Designs and Trade Marks) and C-6/11 (Daiichi Sankyo Company v Comptroller-General of Patents). published on the website of the European Court of Justice (http://curia.europa.eu).

2.4 Paediatric Investigations (EU)

Under Regulation (EC) 1901/2006, the protection conferred by an SPC can be extended by another 6 months in case the patentee has carried out a paediatric investigation plan (PIP), which is now a must for most new drug approvals in Europe. The said extension will also be granted in case the PIP reveals that the respective drug is not suitable for paediatric use, or if the disease for which the drug is intended occurs only in adults.

One issue not yet resolved is the question whether or not an SPC can be granted which has zero or even negative lifetime. In case of the drug Sitagliptin, for example, the time between patent application and market authorization was less than 5 years. The SPC lifetime would thus end about 100 days prior to the end of the patent term. Nonetheless, the applicant, who has also provided a PIP to the European Medicines Agency (EMA), filed requests for SPCs in the different EU member states, probably to obtain the paediatric extension of 6 months in order to receive about 80 days of extended protection. The requests have been treated differently in the different EU member states.

In 2010, the Federal Patent Court (BPatG) referred this case to the ECJ,[13] asking whether or not an SPC can be granted in case the time between patent application and market authorization is less than 5 years. This issue has more than academic importance, because in case an SPC can be granted even if its lifetime is zero, or negative, a paediatric extension thereof can at least lead to an extension of the actual protection by a maximum of six months. Just recently, the Attorney General of the ECJ has issued his opinion, which is indicative, yet non-binding, for the decision of the ECJ, and in which he suggested to allow an SPC even if the period between the date on which the application for a basic patent was lodged and the date of the first authorisation to place the product on the market in the community is less than five years.

Should this opinion be adopted by the ECJ it would make sense to apply for an SPC even in case it's lifetime is negative, at least in case the lifetime of the SPCs ends somewhere in the second half of the twentieth patent year.

Because, despite prior hopes, the development and approval of a new therapeutic antibody is at least as cost-intensive and lengthy as that of a small molecular drug, patent term extensions and SPCs will probably be at least as important for biopharmaceutics as they have been for small molecular drugs.

---

[13] Federal Patent court 15 W (patent) 36/08, case C-125/10 (Merck & Co Inc v Deutsches Patent- und Markenamt), published on the website of the European Court of Justice (http://curia.europa.eu).

3 Exclusivity Privileges Related to Regulatory Procedures

3.1 Data Exclusivity and Market Exclusivity

Another additional protective instrument for drugs including biopharmaceutics is test data exclusivity and/or market exclusivity, as provided by most countries. The use of these terms is not fully consistent in the different legal systems. However, there seems to be consent that, under test data exclusivity, a biosimilar manufacturer is banned from relying on, or referring to, approval data relating to the original drug, in their own approval applications, even when Patent protection of the latter has expired. Test data exclusivity does, however, not prevent a third party from generating their own data—which can become a costly matter.

Market exclusivity, in contrast, defines the term in which a biosimilar manufacturer will not receive an approval from the respective authority, even though he may have requested for such approval and presented the necessary data already, even if data have been taken from the authorization of the innovator.

Both instruments are compliant with Article 39 (3) of the TRIPS agreement (Trade Related Aspects of Intellectual Property Rights), a contract which all members of the World Trade Organization (WTO) have signed.

In the European Union, the so-called $8 + 2 + 1$ formula applies under Regulation (EC) 726/2004, according to which an eight-year data exclusivity term is provided, said term beginning with the market authorization of the original drug, and under the condition that a new indication with significant clinical benefit compared with existing therapies is provided. A two-year market exclusivity is furthermore provided, the latter being extendible by another year in case one or more new therapeutic indications are found in the eight year period. This regulation applies to biopharmaceutics which have been submitted for approval after October 31, 2005. For drugs submitted earlier, a data exclusivity of 10 years applies for applications filed before the EMA, while for national applications or mutual recognition procedures a data exclusivity of 6 years applies, with some countries (Belgium, France, Germany, Italy, Luxembourg, Netherlands, Sweden and United Kingdom) expanding this term to 10 years.

In case a biopharmaceutic is not or no longer protected by a patent or SPC, supplementary paediatric research can be rewarded by a paediatric use marketing authorization (PUMA), which involves 10 years of data exclusivity for the data which have been used to obtain paediatric-use marketing authorisation.

In the US, a data exclusivity of 5 years for small molecular drugs (also called new chemical entities, or NCE), and 3 years for new indications, was adopted under the Hatch-Waxman Act, with add-ons of 6 months for drugs on which the US Food and Drug Administration (FDA) has requested paediatric studies. As regards biopharmaceutics, the Patent Protection and Affordable Care Act, which issued 2010, provides a data exclusivity term of 12 years, with 6 months extension in case of paediatric studies. For the first 4 years of this period (or 4 ½ years in case paediatric exclusivity is granted), the FDA will not review any biosimilar applications which refer to the respective reference biopharmaceutic. Further, the Act provides a so-called biogeneric exclusivity for the party which is the first to obtain an interchangeable license, i.e. the first approved biosimilar (first follower). Said biogeneric exclusivity provides protection against subsequent interchangeable license applications and is thus a market exclusivity which ranges between 12 and 42 months depending on the circumstances (e.g., whether the parties engage in legal action).

3.2 Orphan Drugs

Orphan drugs are drugs which are used to treat rare diseases. In order to provide sufficient incentive to develop such types of drugs, which are commercially less attractive than blockbuster drugs, additional exclusivity instruments have been created.

Under Regulation (EC) 141/2000, a market exclusivity of 10 years can be granted for an orphan drug. According to the regulation, a disease qualifies as an orphan disease in case it is a life-threatening, seriously debilitating or serious and chronic condition, and affects less than 230,000 patients/year, or less than 5 patients per 10,000 persons in the European Community. Further, the orphan disease status is also assigned to some tropical diseases that are mainly found in developing countries. The orphan drug status is however bound to a specific indication, which means that in case the respective drug is already on the market with respect to other indications, off label use may be possible. Further, the market exclusivity term is subject to revisions, and can be cut to 6 years if the conditions required for a rare disease status are no longer met.

In the US, the Orphan Drug Act (ODA) grants particular privileges for drugs addressing disorders affecting fewer than 200,000 patients/year, or less than 7,5 patients per 10,000 persons. In addition to tax incentives and other financial advantages, these privileges encompass a 7 year market exclusivity.

On May 15, 2011, the European Register of designated Orphan Medicinal Products[14] listed 727 registered Orphan Medicinal Products, out of which 70 related to therapeutic antibodies, or antibody mimetics.[15]

Consequently, one strategy companies which have developed a biopharmaceutic against a target for which an earlier biopharmaceutic has already been authorized may want to use is to obtain orphan drug status for an indication not yet covered by the earlier biopharmaceutic.

This was, for example, the case for Nimotuzumab, which is a humanized anti-EGFR IgG, and thus competes with Merck's Erbitux. Nimotuzumab is protected, among others, under US Patent 5,891,996. It has been developed at the Centre of

---

[14] Available under http://ec.europa.eu/health/documents/community-register/html/orphreg.htm.

[15] Storz (2011).

Molecular Immunology in Havana, Cuba, and has orphan drug status, among others, in the EU (treatment of glioma and pancreatic cancer, EU designation number EU/3/04/220 and EU/3/08/550), and in the US (treatment of glioma).

Another example is LFB-R603 developed by LFB of France, which is a chimeric anti CD20 IgG. LFB claims that LFB-R603 exhibits improved antibody dependent antibody-dependent cell-mediated cytotoxicity (ADCC) with respect to its reference antibody, Rituximab. LFB-R603 has obtained orphan drug status for the treatment of chronic lymphocytic leukaemia (CLL) both in the US and in Europe (EU/3/09/699).

Table 2 gives an overview of monopoly rights and exclusivity privileges which biopharmaceutic companies can, and should, make use of.

4 Patent Lifecycle Management

In addition to the discussed legal instruments which provide means for the prolongation of exclusivity granted by a given patent, or which relate to a given market authorization, particular strategies exist to extend efficient patent protection of a given biopharmaceutic, namely by filing a series of subsequent patent applications which reflect the history of said biopharmaceutic, e.g., the discovery of a physiologically attractive target, a specified active compound, new formulation or galenics for such compound, a combination product comprising said compound, use of said compound for a new medical indication or a new dosage regimen for a said biopharmaceutic. Table 3 gives an overview of different embodiments which can be subject of such patent lifecycle management strategy.

By applying such strategy, innovators can substantially extend the effective patent protection for a given biopharmaceutic. Further, this strategy takes into account that, at a time a new target is discovered (and, as such, a patent application for such target should be filed) a concrete drug addressing this target is not yet within sight. Likewise, once a potential drug candidate has been identified, a patent application should again be filed, again even in case said drug is not yet ready for authorisation. An appropriate patent lifecycling strategy thus reflects the different stages of a biopharmaceutic drug development process, plus new embodiments which represent second or higher generation advancements of said biopharmaceutic. The different stages will be discussed in the following.

4.1 New Biological Agents

Although most physiological and signalling processes on the cellular level are today well-understood, new biological agents, e.g., receptors, ligands, enzymes, nucleic acid sequences, and so forth, are still being discovered.

A new biological agent can be subject of a patent application at least as long as a physiological role of said agent is also disclosed, wherein said role has the

Table 2 Overview of monopoly rights and exclusivity privileges

	Requirements	US	Europe	Effects
Monopoly rights				
Patent	Novelty and non-obviousness	20 years	20 years	Patentee, or owner of SPC, can block third parties to make, use, offer to sell, or sell the patented invention. Claims can be enforced before court. In case of compulsory licenses, compensatory payments can be granted in some jurisdictions
PTA	Delays in prosecution caused by the USPTO	10 days	–	–
SPC/PTE	Time consuming approval proceedings	5 years (max)	5 years (max)	–
Paediatric extension of SPC	Paediatric investigation + grated SPC	–	6 months	–
Priority claim	–	1 year	1 year	–
Exclusivity privileges				
Data exclusivity	New drug with full approval	12 years	8 years under 8 + 2 + 1 scheme	Third parties cannot refer to clinical test data lodged by the originator to approve a biosimilar, plus FDA will not review any biosimilar applications for the first 4 years (i.e. aspects similar to market exclusivity
Market exclusivity		–	2 years under 8 + 2 + 1 scheme, plus 1 year if new indication registered in first 8 years	Biosimilar application can be submitted, but approval will not be granted prior to end of the exclusivity term

(continued)

Table 2 (continued)

	Requirements	US	Europe	Effects
Paediatric extension of data exclusivity	Paediatric studies required	6 months	–	See data exclusivity
PUMA	Biopharmaceutic is not protected by patent or SPC, supplementary paediatric research is performed	–	10 years	Exclusivity for data which have been used to obtain paediatric-use marketing authorisation
Generic exclusivity	First follower	12–42 months	–	Protection against subsequent applications for biosimilar approval
Orphan drug exclusivity	Drugs addressing disorders affecting less than 200,000 (US) or 230,000 (EU) patients/year	7 years	10 years	US: market exclusivity plus other incentives; EU: market exclusivity

Table 3 Examples for different embodiments which can be subject of patent lifecycle management strategy

Generation	Protected subject matter	Example patent	Claim language
1	Target	EP0186833B2 (Yeda)	"A monoclonal antibody which [...] binds a cytotoxin having M. W. of 17.000 [...] wherein said cytotoxin exhibits a cytotoxic effect on CHI-sensitised SV-80 cells and is obtainable in a state of enhanced purity by adsorption [...] onto controlled pore glass beads [...]"
2	Active compound	US5731169 (Societe Leb-Tech)	"1. An isolated DNA molecule encoding a receptor for human alpha interferon comprising the amino acid sequence set forth in SEQ ID NO: 2 [...]"
3	Formulations and/or Galenics	EP0652766B1 (Genentech)	"A stable [...] formulation [...]comprising human growth hormone, a buffer providing a pH in the range of 5.5–7, 0.1–1% by weight of a non-ionic surfactant, a neutral salt, and a preservative"
4	Combination products	EP1169059B1 (Aventis)	"A therapeutic pharmaceutical combination, comprising Docetaxel and rhuMAb HER2"
5	Second or higher medical indication	EP0230980B2 (Amgen)	"The use of a human granulocyte colony stimulating factor [...] for [...] a pharmaceutical composition for the recovery of hematopoietic capacity following bone marrow transplantation [...]"
6	New dosage regimen	EP0643965B1 (Abbott)	"Use of nicotinic acid [...] for [...] a sustained release medicament for use in the treatment by oral administration once per day prior to sleep, of hyperlipidaemia [...]"

potential to provide a commercial use of said novel biological agent for physiological intervention, e.g., by means of a pharmaceutical (or a crop protection product), in which case the new biological agent would be called a target. Today, about 100 biological targets are addressed by approved biopharmaceutics,[16] but the number of yet undiscovered targets is probably much higher.

Other potential role comprise the use of said new biological agent as an active compound itself, e.g., as a therapeutic, a food supplement, in plant protection, in industrial processes, or the use of said new biological agent in diagnostics, in drug screening and so forth.

[16] Overington et al. (2006).

Filing a patent application for, e.g., an antibody against a new target is, usually, a safe bet. In case a patent application describes, and claims, a new protein that may play a physiological role in the human body, it is common to also draft a claim related to a theoretical antibody against said protein. Such type of claim is routinely granted in case the target protein is novel and substantially defined, even if the applicant has not produced such an antibody, or provides no data or enablement related to such antibody (see, for example, US decision Noelle versus Lederman,[17] or EPO technical board decision T0542/95).[18] In both cases, the rationale behind this position was that the provision, and correct specification, of a novel protein X enables a skilled person to produce an antibody against said protein. Therefore, it is considered a fair reward for the applicant of protein X to be granted a claim related to a theoretical antibody against said protein.

4.2 Active Compound

Once a given target is prior art, patents can still be obtained with respect to particular compounds interacting with said target. In biopharmaceutics, active compounds can be claimed in different ways, e.g., with reference to a given nucleic acid sequence (encoding nucleic acid sequence in case of a protein compound, or nucleic acid sequence as such in case the compound is a nucleic acid, like an aptamer or antisenseRNA) or amino acid sequence (in case of a protein compound). Other approaches encompass reference to particular binding properties of said compound against a target, to a deposited cell line which produces said compound, or with respect to a process with which the compound can be produced.

It needs to be said that the concept of absolute compound protection (i.e. protection of a new compound without any restriction towards a given purpose), which is established granting practice at least in Europe for small molecular drugs, is subject to serious challenges, at least in the biopharmaceutic field.[19]

4.3 Formulations and/or Galenics

A new formulation for a known biopharmaceutic can as well be made subject of a patent application. Such approach can contribute to keep competitors at bay, particularly in case the new formulation has considerable benefits over formulations for said biopharmaceutic known from the prior art. Drug delivery systems which are specific for a given biopharmaceutic (e.g., as is the case for antisense RNA) fall under this category, too.

---

[17] Decision Noelle v. Lederman, 355 F.3d 1343, 2004 U.S. App. LEXIS 774.

[18] EPO Technical Board decision T0542/95, 1999, EPO Board of appeal decisions database No T0542/95.

[19] Hüttermann and Storz (2010).

4.4 Combination Products

Very often, it turns out that a new combination of two or more pharmaceutical components may have a synergistic effect. In such case, said combination can be made subject of a patent application, in order to monopolize said combination. Particularly in cancer therapy, drug combinations are often used, e.g., a monoclonal antibody plus a cytostatic agent. In case an innovator files a patent application related to such drug combination (which would come at a time when the monoclonal antibody is prior art already) he can block competitors from selling such combination products even in case the patents protecting the individual components have expired.

However, many patent jurisdictions[20] penalise not only direct patent infringement, but also indirect patent infringement (also called contributory infringement). A constellation in which a patent claim encompasses two or more features (e.g., compounds A and B), and a third party supplies only compound A, could be for example considered as indirect infringement under the condition that compound A can only reasonably be used to make the claimed subject matter, for example because the only existing market authorization is for A and B.

A patent related to a pharmaceutical combination which is also subject of a market authorization (while neither of the two individual components is) can thus provide protection even against attempts, by third parties, to sell the individual components, even if the latter are free state of the art.

4.5 Second or Higher Medical Indication

In case a new indication has been found for a drug which is already prior art, such new indication can be made subject of a patent application. Because the respective claims will focus on a method of use, care has to be taken to ensure that such claims do not qualify as a method of treatment, which is unpatentable in many patent jurisdictions, e.g. under the EPC,[21] or largely unenforceable in others, like the United States.[22]

See Table 3 for a suitable claim wording as accepted by the EPO. In the US, "use" is not a claim category as provided by the US Patent Act.[23] Therefore, the

---

[20] e.g., § 10 of the German Patent Act, or 35 USC § 271(c).

[21] Article 54(5) EPC: paragraphs 2 and 3 shall also not exclude the patentability of any substance or composition referred to in paragraph 4 for any specific use in a method referred to in Article 53(c), provided that such use is not comprised in the state of the art.

[22] 35 USC §287(c)(1): [...] the provisions of sections 281, 283, 284, and 285 of this title shall not apply against the medical practitioner or against a related health care entity with respect to such medical activity.

[23] 35 USC 101: "whoever invents or discovers any new and useful process, machine, manufacture, or composition of matter, or any new and useful improvement thereof, may obtain a patent therefore".

claim wording should be as follows: "a process comprising administering a composition comprising Antibody XY to a human in an amount effective for treating a disease Z".

It may seem that the protective effect of such patent is only limited, because of off label use. Under many jurisdictions, the individual prescription of a given drug for a given indication, as effected by a physician, qualifies as exempt from patent protection. Under this privilege, a physician can legally prescribe the biosimilar version of a biopharmaceutic for which compound protection has expired, for treatment of a given disease, even if said indication is subject of second or higher medical indication patent. However, this privilege applies only for prescription on an individual, patient-focused basis.

Under the above circumstances a biosimilar manufacturer can not provide any information with respect to said second or higher medical indication, let alone advertise for said indication. Further, if the second medical indication is the only indication which is clinically relevant, the mere provision of a biosimilar could already qualify as an indirect patent infringement.

4.6 New Dosage Regimen

A new dosage regimen can also be made subject of a patent application, in case it is novel and inventive over a dosage regimen for the same drug, or drug combination, known from the prior art. Again, it may seem that the protective effect of such patent is only limited, because of off label use. However, the mere provision of a biosimilar could as well qualify as an indirect patent infringement in case the claimed dosage regimen is the only one which is clinically relevant, e.g., because the respective market authorization requires such dosage regimen.

5 Conclusion

By well-considered use of patents and related monopoly rights, exclusivity privileges and a suitable patent lifecycle management strategy, biopharmaceutical companies can ensure proper compensation for the efforts and resources invested into research and development of a new biopharmaceutic.

The latter is particularly important to provide sufficient incentive for the development of innovative biopharmaceutics even in the future, because, despite prior hopes it has turned out that the development of a new biopharmaceutic is at least as costly and time consuming as that of a new small molecular drug.

References

DiMasi JA, Hansen RW, Grabowski HG (2003) The price of innovation: new estimates of drug development costs. J Health Econ 22:151–185

Grabowski H, Cockburn I, Long G (2006) The market for follow-on biologics: how will it evolve? Health Aff 25(5):1291–1301

Hüttermann A, Storz U (2010) Monsanto soy bean patent cases—a paradigm shift gathering in case the ECJ takes over patent jurisdiction. LES Nouvelles, Sept 2010, pp 156–159

Overington JP, Al-Lazikani B, Hopkins AL (2006) How many drug targets are there? Nat Rev Drug Discov 5:993–996

Storz U (2011) Intellectual property protection: strategies for antibody inventions. mAbs 3(3):310–317

About the Authors

Dr. Andreas Hübel born in 1964, studied Biology at the University of Hamburg. After graduation, he worked at the Bernhard-Nocht-Institute for Tropical Medicine and received his Ph.D. in 1996 for a thesis in the field of molecular microbiology. He worked as postdoctoral fellow at Harvard Medical School and Washington University School of Medicine before joining the Center for Hygiene at Philipps-University Marburg. He is coauthor of several scientific articles in the fields of molecular neurobiology, molecular parasitology and glycobiology.
Andreas Hübel began his professional career in intellectual property in 2000 and was admitted to bar as German Patent Attorney and as European Patent Attorney in 2005. As chartered European Trademark and Design Attorney he is also admitted to represent clients before the Office of Harmonization for the Internal Market. The main fields of his activity in intellectual property comprise drafting and prosecuting patent applications, attacking or defending patents, and assessing validity of intellectual property rights and Freedom-to-Operate in the areas of pharmaceuticals, biochemistry, biotechnology and adhesives.

Thilo Schmelcher born in 1970, studied Electronics at the University of Karlsruhe (TH) having the major subject of Optical Telecommunication Engineering. There, he also prepared his diploma thesis relating to the usage of the transfer matrix method in resonant tunneling in finite superlatices. Thereafter he co-worked at the newly established chair of optical communication at the University of Mannheim on integrated waveguides. Being interested in intellectual property rights since 1996, Thilo Schmelcher started his IPR career when joining Munich-based patent law firm df-mp in 2002.
In 2005, he was admitted as German Patent Attorney as well as European Trademark and Design Attorney, in 2008 as European Patent Attorney. Until the end of 2005 he was associate of Munich based patent law firm df-mp where he prosecuted oppositions and performed freedom-to-operate studies. Afterwards, he joined the patent department of Telefonaktiebolaget L M Ericsson and

A. Hübel et al., *Biopatent Law: Patent Strategies and Patent Management*,
SpringerBriefs in Biotech Patents, DOI: 10.1007/978-3-642-24846-7,

drove and conducted several strategic projects and multinational nullity and infringements suits.
His main areas are in the fields of telecommunication, automation, medical devices, software related inventions, metrology and cybernetics, as well as optics and lasers. He works as well in the field of Trademarks and Design. He works on prosecution, enforcement and nullification of the respective IPRs.

Dr. Ulrich Storz was born in 1969 in Muenster. He graduated in Biology from the University of Muenster in 1998, where he received his Ph.D. in 2002. He is author and coauthor of several scientific publications in the field of biology and biophysics as well as of several juridical publications in the field of intellectual property. He passed the German Patent Bar Examination in 2005. Since 2005, he has been admitted to practice as European Trademark Attorney at the European Trademark Office (OHIM). In 2006, he was registered in the list of representatives before the European Patent Office. His Main practice areas in the field of Intellectual Property Law include Patent Prosecution, FTO and Patent Infringement, as well as Patent strategies; especially in the Life Science field (i.e. Biotechnology, Biophysics, Biochemistry, microbiology). One of his major fields of interest is Antibody IP. Ulrich Storz is active as a speaker for the congress management company "Forum Institut für Management GmbH", and he organizes the annual Rhineland Biopatent Forum.
Ulrich Storz is the series editor of *SpringerBriefs in Biotech Patents*.

GPSR Compliance

The European Union's (EU) General Product Safety Regulation (GPSR) is a set of rules that requires consumer products to be safe and our obligations to ensure this.

If you have any concerns about our products, you can contact us on ProductSafety@springernature.com

In case Publisher is established outside the EU, the EU authorized representative is:

Springer Nature Customer Service Center GmbH
Europaplatz 3
69115 Heidelberg, Germany

Batch number: 10682325

Printed by Printforce, the Netherlands